Editor
Lorin E. Klistoff, M.A.

Managing Editor
Karen Goldfluss, M.S. Ed.

Editor-in-Chief
Sharon Coan, M.S. Ed.

Cover Artist
Barb Lorseyedi

Art Director
CJae Froshay

Art Manager
Kevin Barnes

Imaging
Rosa C. See

Product Manager
Phil Garcia

Publishers
Rachelle Cracchiolo, M.S. Ed.
Mary Dupuy Smith, M.S. Ed.

Author
Robert W. Smith

Teacher Created Materials, Inc.
6421 Industry Way
Westminster, CA 92683
www.teachercreated.com
ISBN-0-7439-8635-0
©2004 Teacher Created Materials, Inc.
Made in U.S.A.

The classroom teacher may reproduce copies of materials in this book for classroom use only. The reproduction of any part for an entire school or school system is strictly prohibited. No part of this publication may be transmitted, stored, or recorded in any form without written permission from the publisher.

Table of Contents

Introduction 3
Practice 1: Number Sentences/Missing Terms/Boxes and Symbols 4
Practice 2: Number Sentences/Missing Terms/Variables 5
Practice 3: Missing Factors 6
Practice 4: Algebraic Symbols 7
Practice 5: Evaluating Simple Expressions 8
Practice 6: Evaluating Simple Expressions 9
Practice 7: Axioms of Equality (Addition) 10
Practice 8: Axioms of Equality (Subtraction) 11
Practice 9: Axioms of Equality (Multiplication) 12
Practice 10: Axioms of Equality (Division) 13
Practice 11: Solving Equations (Mixed Operations) 14
Practice 12: Solving Equations (Mixed Operations) 15
Practice 13: Solving Two-Step Equations 16
Practice 14: Number Sentences 17
Practice 15: Number Sentences 18
Practice 16: Evaluating Exponents 19
Practice 17: Evaluating Exponents in Expressions 20
Practice 18: Evaluating Exponents with Prime Factors 21
Practice 19: Using Parentheses in Expressions 22
Practice 20: Order of Operations (Four Operations) 23
Practice 21: Order of Operations (Four Operations) 24
Practice 22: Order of Operations (PEMDAS) 25
Practice 23: Order of Operations (PEMDAS) 26
Practice 24: Evaluating Expressions with Variables 27
Practice 25: Evaluating Expressions with Variables 28
Practice 26: Ordering Integers 29
Practice 27: Adding Integers 30
Practice 28: Adding Integers 31
Practice 29: Applying Formulas (Perimeter: Rectangles/Squares) 32
Practice 30: Applying Formulas (Perimeter: Parallelograms/Triangles) 33
Practice 31: Applying Formulas (Area: Rectangles/Squares) 34
Practice 32: Applying Formulas (Area: Parallelograms) 35
Practice 33: Applying Formulas (Circumference) 36
Practice 34: Simple Functions 37
Practice 35: Simple Functions 38
Practice 36: Sequences 39
Test Practice Pages 40
Answer Sheet 46
Answer Key 47

Introduction

The old adage "practice makes perfect" can really hold true for your child and his or her education. The more practice and exposure your child has with concepts being taught in school, the more success he or she is likely to find. For many parents, knowing how to help your children can be frustrating because the resources may not be readily available. As a parent it is also difficult to know where to focus your efforts so that the extra practice your child receives at home supports what he or she is learning in school.

This book has been designed to help parents and teachers reinforce basic skills with their children. *Practice Makes Perfect* reviews basic math skills for children in grade 5. The math focus is on pre-algebra. While it would be impossible to include all concepts taught in grade 5 in this book, the following basic objectives are reinforced through practice exercises. These objectives support math standards established on a district, state, or national level. (Refer to the Table of Contents for the specific objectives of each practice page.)

- missing terms/boxes, symbols, or variables
- missing factors
- algebraic symbols
- evaluating simple expressions
- axioms of equality (addition, subtraction, multiplication, division)
- solving equations (mixed operations)
- solving two-step equations
- number sentences
- evaluating exponents
- using parenthesis in expressions
- order of operations
- evaluating expressions with variables
- ordering and adding integers
- applying formulas (perimeter, area, and circumference)
- simple functions
- sequences

There are 36 practice pages organized sequentially, so children can build their knowledge from more basic skills to higher-level math skills. (*Note:* Have children show all work where computation is necessary to solve a problem.) Following the practice pages are six test practices. These provide children with multiple-choice test items to help prepare them for standardized tests administered in schools. Use the fill-in answer sheet on page 46. To correct the test pages and the practice pages in this book, use the answer key provided on pages 47 and 48.

How to Make the Most of This Book

Here are some useful ideas for optimizing the practice pages in this book:

- Set aside a specific place in your home to work on the practice pages. Keep it neat and tidy with materials on hand.
- Set up a certain time of day to work on the practice pages. This will establish consistency. An alternative is to look for times in your day or week that are less hectic and more conducive to practicing skills.
- Keep all practice sessions with your child positive and constructive. If the mood becomes tense or you and your child are frustrated, set the book aside and look for another time to practice with your child.
- Help with instructions if necessary. If your child is having difficulty understanding what to do or how to get started, work through the first problem with him or her.
- Review the work your child has done. This serves as reinforcement and provides further practice.
- Allow your child to use whatever writing instruments he or she prefers. For example, colored pencils can add variety and pleasure to drill work.
- Pay attention to the areas with which your child has the most difficulty. Provide extra guidance and exercises in those areas. Allowing children to use drawings and manipulatives, such as coins, tiles, game markers, or flash cards, can help them grasp difficult concepts more easily.
- Look for ways to make real-life applications to the skills being reinforced.

Number Sentences/Missing Terms/Boxes and Symbols

Practice 1

Directions: Write the missing number, represented by the square or triangle, on the line below the number sentence.

1. 22 + 13 = □

2. 19 + 53 = △

3. 38 + 79 = □

4. 67 + 44 = □

5. 16 + △ = 33

6. △ + 19 = 44

7. △ + 13 = 50

8. 23 + △ = 67

9. 27 + △ = 56

10. □ + 21 = 60

11. □ + 49 = 76

12. 9 + △ = 77

13. 13 + □ = 90

14. □ + 24 = 63

15. 17 + △ = 59

16. △ + 32 = 81

17. □ + 17 = 51

18. 15 + △ = 93

19. 47 + □ = 218

20. △ + 37 = 123

21. 13 + △ = 100

22. □ + 96 = 119

23. △ + 27 = 119

24. 28 + □ = 113

Practice 2

Directions: Write the missing number, represented by the letter, on the line below the number sentence.

1. 9 + 23 = c

2. 29 + 40 = n

3. 48 + 12 = d

4. 56 + 17 = p

5. 6 + d = 36

6. n + 10 = 45

7. 11 + q = 31

8. c + 10 = 49

9. b + 13 = 50

10. r + 9 = 59

11. d + 17 = 27

12. 8 + s = 37

13. 13 + n = 33

14. r + 6 = 22

15. 25 + c = 85

16. d + 25 = 55

17. 8 + c = 50

18. n + 12 = 36

19. n + 33 = 56

20. 25 + r = 85

21. n + 30 = 80

22. 28 + s = 56

23. 22 + c = 88

24. c + 17 = 97

Missing Factors

Practice 3

Directions: In each number sentence, write the missing factor in the box.

1. 9 x ☐ = 63
2. 7 x ☐ = 56
3. 12 x ☐ = 60

4. 6 x ☐ = 48
5. 6 x ☐ = 72
6. ☐ x 9 = 72

7. ☐ x 12 = 84
8. 11 x ☐ = 88
9. 7 x ☐ = 35

10. ☐ x 11 = 121
11. ☐ x 6 = 60
12. 8 x ☐ = 64

13. 5 x ☐ = 50
14. ☐ x 9 = 108
15. ☐ x 25 = 100

16. 30 x ☐ = 90
17. 25 x ☐ = 150
18. ☐ x 15 = 90

19. 13 x ☐ = 169
20. ☐ x 14 = 56
21. ☐ x 33 = 99

22. 20 x ☐ = 160
23. 20 x ☐ = 200
24. ☐ x 10 = 190

Algebraic Symbols

Practice 4

Reminder

The following are symbols often used in algebra:
- = is equal to
- ≠ is not equal to
- > is greater than
- < is less than
- ( ) parenthesis
- / divided by
- • multiplication dot
- ( )( ) back-to-back parentheses: must multiply

Directions: Do the operational work in each expression and write the answer. The first two are done for you.

1. 9 + 10 > 9 + 7
         19 is greater than 16

2. (7)(6)
         42

3. 4 + 15 < 12 + 12

4. 2 x 9 < 3 x 11

5. (12)(10)

6. 8(7) + 6

7. 18/3 > 5

8. 3(15) – 14

9. 63/7 + 13

10. 9 x 9 > 8 • 9

11. 45 > (3)(12)

12. 27 – 9 < 35

13. 8 • 6 < 77

14. 96/12 = 2 x 4

15. 44 ≠ 10 • 4

16. (9)(6) ≠ 67

17. (13)(8) > 4 x 16

18. 27/3 = 3(3)

19. (19)(2) < 40

20. 44 – 17 = 9 x 3

© Teacher Created Materials, Inc.   #8635 Practice Makes Perfect: Pre-Algebra

Evaluating Simple Expressions

Practice 5

> **Reminder**
> - When you *evaluate* an expression, you compute its numerical value.
> - Multiply a number next to a parenthesis times the number in the parenthesis.
>   6(8) = 48
> - Multiply the numbers inside back-to-back parentheses.
>   (9)(8) = 72

Directions: Evaluate these expressions. The first one is done for you.

1. 4(11) − 7
   44 − 7 = 37
   __37__

2. (7)(12) − 13
   ____

3. (5)(10) − 11
   ____

4. 7(8) − 18
   ____

5. 5(12) − 7(3)
   ____

6. 4(20) − 22
   ____

7. 8 + 13 − 11
   ____

8. 25 − 5(3)
   ____

9. 17 + 12 − 9
   ____

10. 9 × 8 − 2
    ____

11. 48/12 + 9(3)
    ____

12. 11 − 9 + 16
    ____

13. (3)(3) + 7
    ____

14. 7(20) − 13
    ____

15. (25)(1)(2)
    ____

16. 69 − 29 − 4
    ____

17. 56/7 + 16
    ____

18. 4 + 5 − 7
    ____

Evaluating Simple Expressions

Practice 6

> **Reminder**
> - To *evaluate* an expression, you compute its numerical value.
> - Multiply a number next to a parenthesis times the number in the parenthesis.
> $$6(4) = 24$$
> - Multiply the numbers inside back-to-back parentheses.
> $$(7)(8) = 56$$

Directions: Evaluate these expressions.

1. 7(25) − 40

2. (15)(3)(2)

3. 5(18) − 9

4. 9(20) − 13

5. 7(50) − 19

6. (7)(19) − 27

7. 37 + 20 − 17

8. (6)(15) + 32

9. 90 − 17 + 31

10. 8(4) − 31

11. (12)(9) − 7

12. (7)(40) − 88

13. 67 + 13 − 19

14. (7)(4) − 14

15. 8(8) − 8

16. 15 x 9 x 0

17. 49 − 7 + 15

18. (12)(1)(4)

19. 9 x 3 − 14

20. (25)(12) − 44

21. 9(21) − 6

Axioms of Equality (Addition)

Practice 7

> **Reminder**
> Any number added to one side of an equation can be added to the other side.
> $$n - 7 = 27$$
> $$\underline{+7} \quad \underline{+7}$$
> $$n = 34$$

Directions: Solve these equations by adding the same number to both sides of the equation.

1. $n - 9 = 11$
   $\underline{+9} \quad \underline{+9}$
   $n = 20$

2. $n - 6 = 30$

3. $c - 5 = 25$

4. $b - 9 = 16$

5. $n - 11 = 9$

6. $g - 13 = 17$

7. $a - 15 = 20$

8. $p - 18 = 12$

9. $n - 25 = 25$

10. $b - 20 = 10$

11. $v - 19 = 21$

12. $m - 14 = 26$

13. $t - 17 = 17$

14. $c - 8 = 12$

15. $a - 13 = 37$

16. $n - 11 = 22$

17. $d - 21 = 19$

18. $g - 20 = 30$

Practice 8

> **Reminder**
> Any number subtracted from one side of an equation must be subtracted from the other side.
>
> $$\begin{aligned} a + 18 &= 28 \\ -18 & -18 \\ \hline a &= 10 \end{aligned}$$

Directions: Solve these equations by subtracting the same number from both sides of the equation. The first one is done for you.

1. $n + 19 = 29$
   $-19\ -19$
   $n = 10$

2. $a + 13 = 23$

3. $c + 25 = 35$

4. $d + 11 = 30$

5. $n + 29 = 31$

6. $c + 13 = 20$

7. $d + 8 = 41$

8. $m + 15 = 29$

9. $t + 22 = 44$

10. $a + 17 = 61$

11. $s + 19 = 59$

12. $x + 18 = 36$

13. $p + 12 = 22$

14. $n + 38 = 61$

15. $a + 22 = 88$

16. $t + 25 = 100$

17. $x + 17 = 51$

18. $n + 19 = 71$

Axioms of Equality (Multiplication)

Practice 9

> **Reminder**
> Any number multiplied by one side of an equation must be multiplied by the other side.
>
> Step 1: $\frac{n}{8} = 7$ Step 2: $8 \times \frac{n}{8} = 8 \times 7$ Step 3: $n = 56$
>
> Check: $\frac{56}{8} = 7$

Directions: Solve these equations by multiplying the same number by both sides of the equation. The first one is done for you.

1. $\frac{n}{4} = 10$

   $4 \times \frac{n}{4} = 4 \times 10$

   $n = 40$

2. $\frac{c}{7} = 6$

   $7 \times \frac{c}{7} = 7 \times 6$

   $c = $ _____

3. $\frac{p}{7} = 5$

   _____

4. $\frac{d}{6} = 8$

5. $\frac{t}{6} = 10$

6. $\frac{n}{9} = 6$

7. $\frac{a}{5} = 12$

8. $\frac{c}{9} = 11$

9. $\frac{r}{9} = 7$

Directions: Use the short cut to do these problems. The first one is done for you.

10. $\frac{d}{8} = 12$

    $8 \times 12 = 96$

    $d = 96$

11. $\frac{m}{5} = 6$

12. $\frac{t}{9} = 6$

13. $\frac{c}{3} = 13$

14. $\frac{n}{4} = 20$

15. $\frac{n}{4} = 25$

#8635 Practice Makes Perfect: Pre-Algebra © Teacher Created Materials, Inc.

Axioms of Equality (Division)

Practice 10

> **Reminder**
> Any number divided into one side of an equation must be divided into the other side.
>
> Step 1: $n \times 6 = 24$ Step 2: $n \times \frac{6}{6} = \frac{24}{6}$ Step 3: $n = 4$

Directions: Solve these equations by dividing the same number into both sides of the equation. The first one is done for you.

1. $n \times 8 = 32$

    $n \times \frac{8}{8} = \frac{32}{8}$

    $n = 4$

2. $c \times 5 = 50$

    _____

3. $a \times 9 = 36$

    _____

4. $n \times 12 = 72$

    _____

5. $s \times 5 = 45$

    _____

6. $t \times 9 = 81$

    _____

7. $n \times 3 = 27$

    _____

8. $p \times 12 = 84$

    _____

9. $w \times 8 = 80$

    _____

10. $c \times 9 = 63$

    _____

11. $z \times 11 = 44$

    _____

12. $b \times 9 = 27$

    _____

13. $m \times 6 = 36$

    _____

14. $d \times 7 = 49$

    _____

15. $d \times 8 = 88$

    _____

16. $r \times 6 = 18$

    _____

17. $9 \times c = 54$

    _____

18. $7 \times c = 28$

    _____

Solving Equations (Mixed Operations)

Practice 11

Directions: Solve these equations by using the axioms of equality.

1. $n + 11 = 25$
2. $d \times 10 = 50$
3. $q + 23 = 31$

4. $c - 19 = 22$
5. $m \times 7 = 28$
6. $n \times 9 = 72$

7. $\dfrac{v}{9} = 12$
8. $\dfrac{n}{7} = 8$
9. $n - 6 = 14$

10. $t + 13 = 36$
11. $n \times 8 = 48$
12. $a - 17 = 40$

13. $d \times 12 = 48$
14. $c \times 7 = 49$
15. $\dfrac{d}{6} = 7$

16. $\dfrac{n}{11} = 5$
17. $t - 15 = 29$
18. $s + 24 = 48$

19. $s \times 7 = 77$
20. $d \times 13 = 39$
21. $w - 19 = 44$

22. $t + 24 = 72$
23. $\dfrac{b}{6} = 8$
24. $\dfrac{a}{12} = 12$

Solving Equations (Mixed Operations)

Practice 12

Directions: Solve these equations by using the axioms of equality.

1. $n - 16 = 41$
2. $z \times 9 = 54$
3. $s + 17 = 46$

4. $d - 27 = 53$
5. $t \times 8 = 56$
6. $b + 34 = 91$

7. $\dfrac{n}{6} = 6$
8. $\dfrac{r}{12} = 11$
9. $t - 29 = 56$

10. $v + 19 = 39$
11. $c \times 7 = 84$
12. $b - 33 = 71$

13. $\dfrac{c}{7} = 9$
14. $\dfrac{t}{12} = 6$
15. $v - 25 = 56$

16. $\dfrac{n}{9} = 9$
17. $13 + n = 23$
18. $17 + c = 30$

19. $\dfrac{n}{7} = 4$
20. $q \times 15 = 60$
21. $p + 29 = 91$

22. $r - 26 = 54$
23. $\dfrac{d}{11} = 9$
24. $7 \times n = 77$

Solving Two-Step Equations

Practice 13

Directions: Solve these equations. The first one is done for you.

1. $n + 15 = 8 + 12$
   $n + 15 = 20$
   $\underline{-15 \quad -15}$
   $n = \quad\ \ 5$

2. $p + 8 = 15 + 10$

3. $a - 9 = 7 + 6$

4. $r + 16 = 20 + 3$

5. $n - 19 = 2 + 7$

6. $c + 12 = 13 + 15$

7. $b \times 3 = 7 + 20$

8. $4 \times c = 29 + 3$

9. $q + 16 = 4 + 24$

10. $19 + c = 21 + 14$

11. $9 \times r = 30 + 6$

12. $b \times 8 = 12 \times 6$

13. $18 + t = 30 - 4$

14. $c \times 5 = 70 + 5$

15. $23 + r = 40 - 16$

16. $c + 19 = 18 + 21$

17. $12 + n = 28 - 3$

18. $10 \times r = 85 - 15$

Practice 14

Directions: Fill in the space with the number that makes each number sentence true. The first one is done for you.

1. 9 + 9 = __14__ + 4
   18 = 14 + 4

2. 12 + 20 = 7 + _____

3. 28 – 7 = _____ + 6

4. 15 + 11 = _____ + 19

5. 10 x 3 = 5 x _____

6. _____ + 8 = 8 x 3

7. 17 + 12 = _____ + 6

8. 9 + 39 = 4 x _____

9. 33 – 12 = 7 x _____

10. _____ + 10 = 46 – 15

11. 27 + 8 = _____ x 5

12. 19 + 19 = _____ + 26

13. 25 + 35 = 10 x _____

14. 16 – 9 = 8 – _____

15. 26 – 7 = 31 – _____

16. _____ x 12 = 31 + 29

17. 18 + 54 = _____ + 23

18. 16 x 3 = 4 x _____

19. 50 – 25 = 5 x _____

20. 43 – 14 = 7 + _____

21. 56 – 18 = _____ + 19

22. 78 + 3 = _____ – 43

23. 70 + 14 = _____ x 12

24. _____ – 17 = 22 + 9

25. 55 ÷ 11 = _____ + 2

26. 48 – 14 = _____ + 17

27. 8 x 6 = _____ + 27

28. _____ – 31 = 9 + 26

29. _____ x 12 = 44 + 52

30. _____ + 2 = 16 – 13

Number Sentences

Practice 15

Directions: Fill in the space with the number that makes each number sentence true. The first one is done for you.

1. 7 + 4 + 9 = <u>26</u> – 6 20 = 26 – 6	2. 13 + 7 – 5 = 6 + _____	3. 8 + 9 – 1 = 8 x _____
4. _____ + 9 – 12 = 19 – 2	5. 14 + 7 – 0 = _____ x 3	6. 15 – 5 + 14 = _____ x 4
7. 28 – 14 + 12 = 9 + _____	8. _____ + 13 = 39 + 6 – 4	9. 29 + 27 – 13 = _____ – 8
10. 8 + 3 – 5 = 5 + _____	11. 50 – 9 – 25 = _____ + 6	12. 20 – 9 – 4 = _____ + 1
13. _____ x 13 = 50 + 2	14. 22 + 5 + 11 = 12 + _____	15. 38 + 3 – 11 = 5 x _____
16. _____ + 12 – 8 = 24 – 7	17. 5 + 5 + 5 = _____ x 5	18. 8 + 8 + 8 = _____ x 3
19. 25 – 6 + 4 = _____ + 6	20. 25 – 5 – 5 = 5 x _____	21. 48 – 8 – 8 = _____ x 8
22. 50 – 10 + 5 = _____ x 9	23. _____ – 6 + 9 = 5 + 8	24. _____ + 20 – 7 = 3 x 12
25. 80 – 57 = 9 + _____	26. _____ – 3 – 2 = 8 + 8	27. 27 – 15 – 6 = _____ + 1
28. 40 – 20 – 10 = _____ + 3	29. _____ + 18 – 5 = 22 – 2	30. _____ + 16 – 8 = 3 x 8

Practice 16

> **Reminder**
> - To determine the value of a term with an exponent of 2, multiply the base times itself.
> $$6^2 = 6 \times 6 = 36$$
> - To determine the value of a term with an exponent of 3, multiply the base times itself. Multiply that answer times the base again.
> $$4^3 = 4 \times 4 \times 4 = 16 \times 4 = 64$$

Directions: Evaluate these expressions by determining the value of the exponents. The first one is done for you.

1. 6^2
    $6 \times 6 = 36$
    $\underline{\quad 36 \quad}$

2. 3^2
    $\underline{\qquad}$

3. 5^2
    $\underline{\qquad}$

4. 2^2
    $\underline{\qquad}$

5. 10^2
    $\underline{\qquad}$

6. 11^2
    $\underline{\qquad}$

7. 5^3
    $\underline{\qquad}$

8. 6^3
    $\underline{\qquad}$

9. 9^2
    $\underline{\qquad}$

10. 8^3
    $\underline{\qquad}$

11. 7^2
    $\underline{\qquad}$

12. 10^3
    $\underline{\qquad}$

13. 7^3
    $\underline{\qquad}$

14. 13^2
    $\underline{\qquad}$

15. 2^3
    $\underline{\qquad}$

16. 3^4
    $\underline{\qquad}$

17. 2^4
    $\underline{\qquad}$

18. 9^3
    $\underline{\qquad}$

19. 5^3
    $\underline{\qquad}$

20. 2^5
    $\underline{\qquad}$

21. 14^2
    $\underline{\qquad}$

22. 1^3
    $\underline{\qquad}$

23. 15^2
    $\underline{\qquad}$

24. 20^2
    $\underline{\qquad}$

25. 30^2
    $\underline{\qquad}$

26. 40^2
    $\underline{\qquad}$

27. 50^2
    $\underline{\qquad}$

Evaluating Exponents in Expressions

Practice 17

Directions: Evaluate these expressions by determining the value of the exponents and completing the operations. The first one is done for you.

1. $2^2 + 9$
   $2 \times 2 = 4$
   $4 + 9 = 13$
   $\underline{13}$

2. $4^2 - 8$
   $\underline{}$

3. $3^2 + 13$
   $\underline{}$

4. $5^2 - 19$
   $\underline{}$

5. $4^2 - 14$
   $\underline{}$

6. $3^3 - 20$
   $\underline{}$

7. $2^3 - 4$
   $\underline{}$

8. $7^2 + 11$
   $\underline{}$

9. $3^3 - 17$
   $\underline{}$

10. $8^2 - 3$
    $\underline{}$

11. $3^2 - 2^2$
    $\underline{}$

12. $7^2 + 15$
    $\underline{}$

13. $2^3 - 2^2$
    $\underline{}$

14. $2^2 + 3^2$
    $\underline{}$

15. $7^2 + 2^2$
    $\underline{}$

16. $5^2 - 2^3$
    $\underline{}$

17. $7^2 + 3^2$
    $\underline{}$

18. $5^2 + 25$
    $\underline{}$

Evaluating Exponents with Prime Factors

Practice 18

Directions: Evaluate these expressions by determining the value of the exponents and completing the operations. The first one is done for you.

1. $2^2 \times 3^2$
   $2 \times 2 = 4 \quad 3 \times 3 = 9$
   $4 \times 9 = 36$
   __36__

2. $2^2 \times 5$
   _____

3. $3^2 \times 7$
   _____

4. $3^2 \times 5^2$

   _____

5. $2^2 \times 11$

   _____

6. $3^3 \times 13$

   _____

7. $3^3 \times 5$

   _____

8. $2^2 \times 7^2$

   _____

9. $5^3 \times 7$

   _____

10. $3^2 \times 7^2$

    _____

11. $11^2 \times 13$

    _____

12. $7^2 \times 11$

    _____

13. $2^3 \times 5^2$

    _____

14. $2^3 \times 7^2$

    _____

15. $7^2 \times 17$

    _____

16. $5^2 \times 11$

    _____

17. $2^3 \times 5 \times 11$

    _____

18. $2^2 \times 5 \times 7$

    _____

© Teacher Created Materials, Inc.

Using Parentheses in Expressions

Practice 19

> **Reminder**
> Always do the operations within the parentheses first when evaluating expressions.
> 22 + (10 − 5) =
> 10 − 5 = 5
> 22 + 5 = 27
> 27

Directions: Evaluate these expressions. Be sure to do the work in the parentheses first. The first problem is done for you.

1. 19 − (7 + 4)
   7 + 4 = 11
   19 − 11 = 8
   __8__

2. (17 − 6) + 11

   _____

3. 31 − (15 + 7)

   _____

4. 29 − (15 + 6)

   _____

5. 8 + (19 − 8)

   _____

6. 17 − (11 − 7)

   _____

7. (6 + 23) − 25

   _____

8. (13 + 12) − 7

   _____

9. 29 − (15 − 4)

   _____

10. (17 + 6) − 8

    _____

11. (34 + 7) − (12 + 4)

    _____

12. (28 − 9) + (7 − 5)

    _____

13. (38 − 15) + (25 + 6)

    _____

14. (50 − 25) + (15 × 2)

    _____

15. (11 × 3) + (9 × 2)

    _____

16. (4 × 7) − (8 × 3)

    _____

17. (17 + 2) + (13 × 1)

    _____

18. (12 × 12) − (11 × 10)

    _____

Practice 20

Reminder

Multiply and divide in order from left to right before you add or subtract.

8 x 3 + 2 x 5
8 x 3 = 24   2 x 5 = 10
24 + 10 = 34
34

Directions: Evaluate these expressions. Be sure to do all multiplication and division operations first. The first problem is done for you.

1. 12 ÷ 4 + 8 x 2
   3 + 16 = 19

   _19_

2. 18 ÷ 6 + 4

   _____

3. 8 x 6 ÷ 4 + 9

   _____

4. 6 + 6 x 4 – 3

   _____

5. 13 + 7 x 4

   _____

6. 9 + 5 x 9

   _____

7. 9 x 4 + 4 x 8

   _____

8. 17 + 6 x 7

   _____

9. 8 + 9 – 2 x 6

   _____

10. 13 – 6 + 7 x 8

    _____

11. 19 – 2 x 6 + 7

    _____

12. 36 ÷ 9 + 7 x 7

    _____

13. 7 + 10 x 5 – 11

    _____

14. 9/3 + 16/4

    _____

15. 72 ÷ 12 x 2 – 11

    _____

16. 36/4 + 9 x 2

    _____

17. 12 x 5 ÷ 6 – 3

    _____

18. 11 x 10 – 5 x 9

    _____

19. 56 ÷ 8 + 7 x 7

    _____

20. 64 ÷ 8 – 3 x 2

    _____

21. 9 x 2 x 1 – 17

    _____

Order of Operations (Four Operations)

Practice 21

Directions: Evaluate these expressions. Be sure to do all multiplication and division operations first. The first problem is done for you.

1. 16 ÷ 8 + 3 x 4
   2 + 12 = 14

   _14_

2. 81 ÷ 9 + 4 x 3

   _____

3. 12 x 3 ÷ 9 + 7

   _____

4. 8 + 8 x 7 − 5

   _____

5. 19 + 8 x 9 − 6

   _____

6. 12 + 6 x 9 − 13

   _____

7. 7 x 4 + 5 x 12

   _____

8. 23 − 2 x 7 + 13

   _____

9. 7 + 11 − 3 x 5

   _____

10. 18 − 4 + 6 x 9

    _____

11. 30 − 3 x 9 + 1

    _____

12. 72 ÷ 9 + 8 x 3

    _____

13. 9 + 11 x 6 − 21

    _____

14. 22 − 3 x 4 + 14

    _____

15. 63 ÷ 9 x 2 − 13

    _____

16. 29 + 3 x 9 − 16

    _____

17. 8 x 9 ÷ 6 − 7

    _____

18. 13 x 10 − 9 x 9

    _____

19. 81/9 − 64/8

    _____

20. 36 ÷ 4 x 3 − 21

    _____

21. 24/6 − 18/6

    _____

22. 17 − 2 x 8 + 11

    _____

23. 28/7 + 3 x 5

    _____

24. 29 + 2 x 5 − 38

    _____

#8635 Practice Makes Perfect: Pre-Algebra © Teacher Created Materials, Inc.

Order of Operations (PEMDAS)

Practice 22

Reminder

Evaluate expressions in this order: PEMDAS

1. <u>P</u>arentheses: Do these operations first
2. <u>E</u>xponents: Find these values next
3. <u>M</u>ultiply and <u>D</u>ivide: In order from left to right
4. <u>A</u>dd and <u>S</u>ubtract: In order from left to right

Directions: Evaluate these expressions. Be sure to follow the order of operations listed above. The first problem is done for you.

1. $4^2 + 9 - (3 \times 5)$
   $4^2 + 9 - 15$
   $16 + 9 - 15 = 10$
   __10__

2. $6^2 - (9 \times 2) + 12$

   _____

3. $10 + (8 \times 3) - 3^2$

   _____

4. $(8 \times 8) - 4^3 + 1$

   _____

5. $8^2 + (4 \times 5) - 21$

   _____

6. $(9 \times 5) - 2^3 + 16$

   _____

7. $15 \div 5 + 7 - 2^2$

   _____

8. $(9 + 11) - 3^2 + 7$

   _____

9. $(6 \times 5) - 14 - 4^2$

   _____

10. $12^2 - 9 \times 12 - 4^2$

    _____

11. $13^2 - (11 \times 9) + 16$

    _____

12. $13 + (6 \times 8) - 5^2$

    _____

13. $17 - (9 + 8) + 2^3$

    _____

14. $9^2 - (27 + 13) - 6^2$

    _____

15. $44 - 7 \times 6 + 4^2$

    _____

Order of Operations (PEMDAS)

Practice 23

Reminder

Evaluate expressions in this order: PEMDAS

1. <u>P</u>arentheses: Do these operations first
2. <u>E</u>xponents: Find these values next
3. <u>M</u>ultiply and <u>D</u>ivide: In order from left to right
4. <u>A</u>dd and <u>S</u>ubtract: In order from left to right

Directions: Evaluate these expressions. Be sure to follow the order of operations listed above. The first problem is done for you.

1. $5^2 + 10 - (6 \times 5)$
$5^2 + 10 - 30$
$25 + 10 - 30 = 5$
    5

2. $7^2 - (7 \times 6) + 4$

3. $15 + (18 \div 3) - 2^3$

4. $(9 \times 5) \div 3 + 1^2$

5. $6^2 + (6 + 8) - 17$

6. $(32 + 32) - 4^3 + 1$

7. $(21 \div 7) + 6 - 2^2$

8. $(19 + 21) - 5^2 + 9$

9. $(7 \times 9) - 3^3 + 2^2$

10. $12^2 - (12 \times 10) + 2^2$

11. $20^2 + (12 \times 5) - 120$

12. $19 + (8 \times 7) + 7^2$

13. $25 - (5 + 9) + 2^4$

14. $11^2 - (16 + 34) - 8^2$

15. $82 - (6 \times 9)$

Evaluating Expressions with Variables

Practice 24

Directions: Evaluate these expressions. The first problem is done for you.

| $a = 2$ | $b = 3$ | $n = 6$ |

1. $n + 11$
  $6 + 11 = 17$
  ___17___

2. $4b - 11$
  _____

3. $3n - 7$
  _____

4. $7n - 19$

  _____

5. $(8)(b) - 23$

  _____

6. $n + a - 5$

  _____

7. $7(n) - 18$

  _____

8. $8b - n$

  _____

9. $8(n) - 2(b)$

  _____

10. $25 - 2n$

  _____

11. $(7)(8) - 4n$

  _____

12. $18 - 3b$

  _____

13. $(10)(6) - 7n$

  _____

14. $n^2 - b^2 + 5$

  _____

15. $n \cdot b + a$

  _____

16. $\dfrac{n}{a} + b - 4$

  _____

17. $\dfrac{n}{b} - 2$

  _____

18. $36 - n^2 + b^2$

  _____

Evaluating Expressions with Variables

Practice 25

Directions: Evaluate these expressions. The first problem is done for you.

| $n = 12$ | $r = 5$ | $t = 8$ |

1. $t - r + 2$
   $8 - 5 + 2 = 5$
   __5__

2. $3r - 13 + t$
   _____

3. $4n - 45 + t$
   _____

4. $7t - 10r + 16$
   _____

5. $10t - 2n + 8$
   _____

6. $n + r - 7$
   _____

7. $7(t) - (9)r + 3$
   _____

8. $12n - 4(t) + 3r$
   _____

9. $7(t) + 3(r) - 9$
   _____

10. $49 - 4(n) - 1$
    _____

11. $t^2 - r^2 + 18$
    _____

12. $(90 - 5n) + n^2$
    _____

13. $(r + t) + (n + 7)$
    _____

14. $(9^2 - t^2) + r$
    _____

15. $(t \cdot r) - 8$
    _____

16. $\frac{n}{2} + 18 - t$
    _____

17. $(t + n) - (r + 9)$
    _____

18. $(48 - r^2) + n^2$
    _____

Practice 26

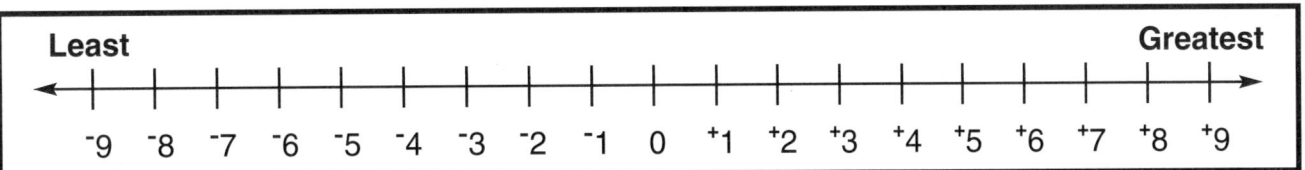

Directions: Order the integers in each set below from least to greatest. Use the number line to help you. The first one is done for you.

1. 0, ⁺1, ⁻2
    ⁻2, 0, ⁺1

2. ⁻6, ⁺4, ⁻5

3. ⁻9, ⁺6, 0

4. ⁻3, ⁻4, ⁻9,

5. ⁻8, ⁺8, ⁻6

6. ⁺7, ⁻6, 0

7. ⁻1, ⁺1, ⁻5

8. ⁺9, ⁻7, 0

9. 0, ⁻9, ⁺6

10. ⁻6, ⁺3, ⁻1

11. ⁻12, ⁺5, ⁻6

12. ⁻7, ⁻5, ⁺6

13. ⁻6, ⁻10, ⁻3

14. ⁻5, ⁻9, 0

15. ⁻8, ⁻5, ⁻13, 0

16. ⁺10, ⁺8, ⁻6, ⁻7

17. ⁻6, ⁻20, ⁻14, ⁺7

18. ⁺11, ⁻12, 0, ⁻7

19. ⁻16, ⁻7, ⁺7, ⁻21

20. ⁺12, ⁻12, ⁻15, 0

21. ⁻5, ⁻9, ⁻4, ⁻17, 0

22. ⁻5, ⁺5, ⁻6, ⁺7, ⁻1

Adding Integers

Practice 27

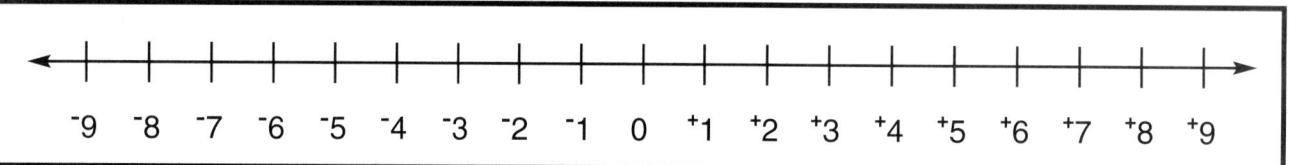

Directions: Add the integers in each problem below. Use the number line to help you. The first one is done for you.

1. $^+8$
   $+\ ^-6$
   $\overline{^+2}$

2. $^+6$
   $+\ ^-9$

3. $^-8$
   $+\ ^-9$

4. $^-9$
   $+\ ^-9$

5. $^-12$
   $+\ ^+10$

6. $^-15$
   $+\ ^+12$

7. $^-19$
   $+\ ^+16$

8. $^-20$
   $+\ ^-12$

9. $^+17$
   $+\ ^-15$

10. $^-14$
    $+\ ^+19$

11. $^+25$
    $+\ ^-29$

12. $^-13$
    $+\ ^+21$

13. $^-22$
    $+\ ^-14$

14. $^+32$
    $+\ ^-31$

15. $^-16$
    $+\ ^-8$

16. $^+13$
    $+\ ^-23$

17. $^-15$
    $+\ ^+22$

18. $^-19$
    $+\ ^+10$

19. $^-16$
    $+\ ^+7$

20. $^+18$
    $+\ ^-11$

21. $^-30$
    $+\ ^-8$

22. $^-25$
    $+\ ^+9$

23. $^-24$
    $+\ ^+5$

24. $^+17$
    $+\ ^-3$

25. $^+40$
    $+\ ^-38$

26. $^-19$
    $+\ ^-8$

27. $^-10$
    $+\ ^+14$

#8635 Practice Makes Perfect: Pre-Algebra © Teacher Created Materials, Inc.

Practice 28

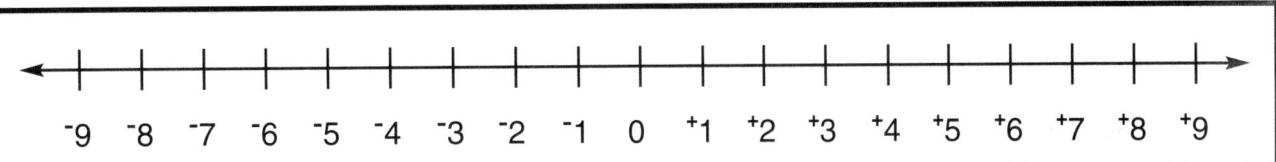

Directions: Add the integers in each problem below. Use the number line to help you. The first one is done for you.

1. ⁺9 + ⁻8 = __⁺1__

2. ⁻9 + ⁻6 = _____

3. ⁻6 + ⁻7 = _____

4. ⁻2 + ⁻8 = _____

5. ⁻9 + ⁻10 = _____

6. ⁻14 + ⁻7 = _____

7. ⁻10 + ⁺16 = _____

8. ⁺20 + ⁻10 = _____

9. ⁻8 + ⁻18 = _____

10. ⁻9 + ⁻6 = _____

11. ⁺15 + ⁻18 = _____

12. ⁻6 + ⁺14 = _____

13. ⁺14 + ⁻19 = _____

14. ⁻7 + ⁻13 = _____

15. ⁻19 + ⁻19 = _____

16. ⁻6 + ⁻17 = _____

17. ⁻4 + ⁻16 = _____

18. ⁻5 + ⁺26 = _____

19. ⁻7 + ⁺36 = _____

20. ⁻3 + ⁺11 = _____

21. ⁺12 + ⁻17 = _____

22. ⁺15 + ⁻30 = _____

Applying Formulas (Perimeter: Rectangles/Squares)

Practice 29

> **Reminder**
> - To compute the perimeter of a rectangle, add two adjoining sides and multiply the sum by 2.
> $$P = (l + w) \times 2$$
> - To compute the perimeter of a square, multiply the length of one side by 4.
> $$P = 4s$$

Directions: Compute the perimeter of each square or rectangle.

1. 6 cm / 4 cm
   P = _____

2. 9 cm / 4 cm
   P = _____

3. 6 in. (square)
   P = _____

4. 17 in. / 12 in.
   P = _____

5. 15 mm / 11 mm
   P = _____

6. 21 mm (square)
   P = _____

7. 16 cm / 12 cm
   P = _____

8. 26 in. (square)
   P = _____

9. 33 m (square)
   P = _____

10. 21 in. / 30 in.
    P = _____

Applying Formulas (Perimeter: Parallelograms/Triangles)

Practice 30

Reminder
- To compute the perimeter of a parallelogram, add two adjoining sides and multiply the sum by 2.
$$P = (l + w) \times 2$$
- To compute the perimeter of a triangle, add the length of the three sides.

Directions: Compute the perimeter of each parallelogram or triangle.

1.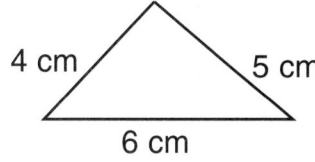

   P = _____

2.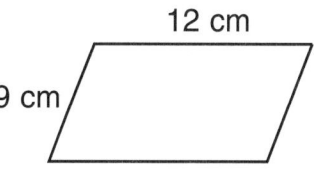

   P = _____

3.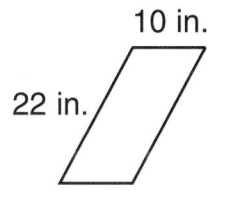

   P = _____

4.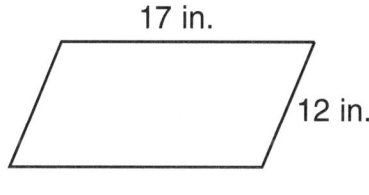

   P = _____

5.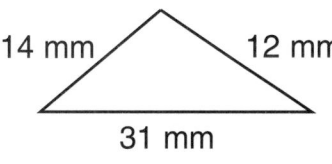

   P = _____

6.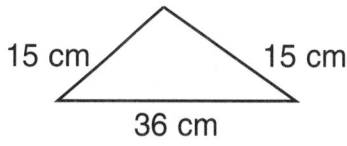

   P = _____

7.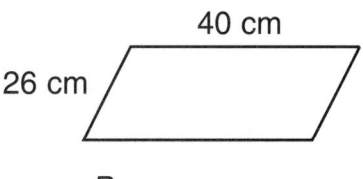

   P = _____

8. 63 in. / 19 in.

   P = _____

9.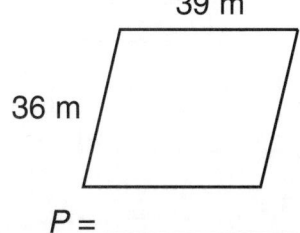

   P = _____

10.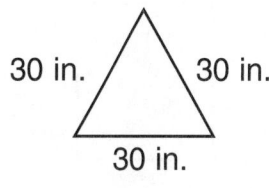

    P = _____

Applying Formulas (Area: Rectangles/Squares)

Practice 31

> **Reminder**
> To compute the area of a rectangle or a square, multiply the length times the width.
> A = l x w

Directions: Compute the area of each rectangle or square.

1.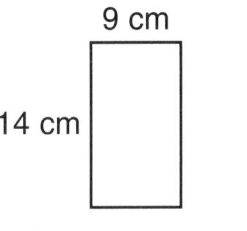
   A = _____

2.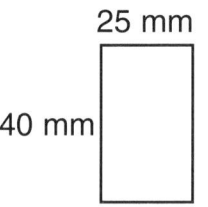
   A = _____

3.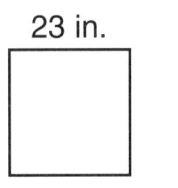
   A = _____

4.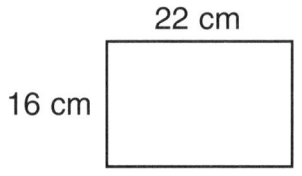
   A = _____

5.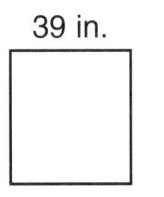
   P = _____

6.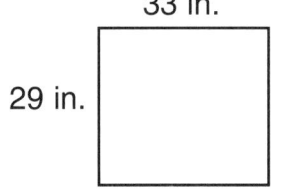
   A = _____

7.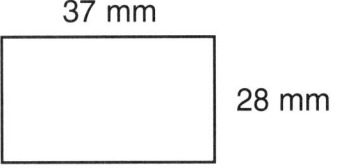
   A = _____

8.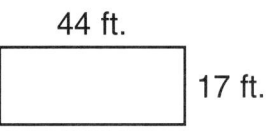
   A = _____

9.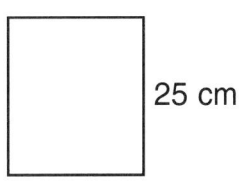
   A = _____

10.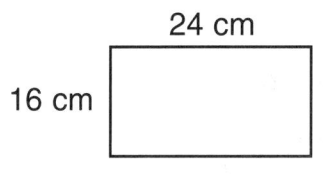
    A = _____

Applying Formulas (Area: Parallelograms)

Practice 32

Reminder
To compute the area of a parallelogram, multiply the base times the height.

A = b x h
A = 6 x 4 = 24 cm

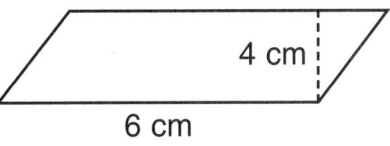

Directions: Compute the area of each parallelogram.

1.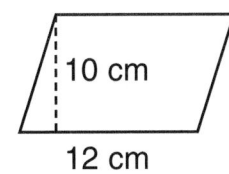

   A = _____

2.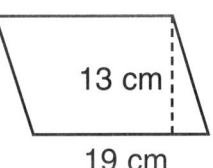

   A = _____

3.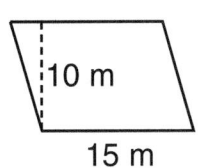

   A = _____

4.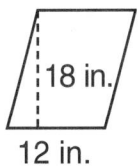

   A = _____

5.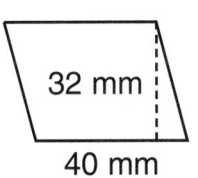

   A = _____

6.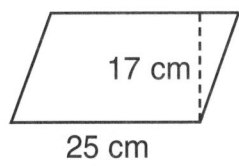

   A = _____

7.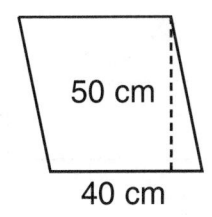

   A = _____

8.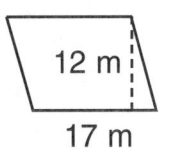

   A = _____

9.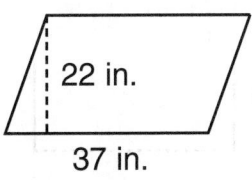

   A = _____

10.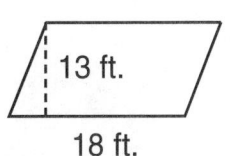

    A = _____

Applying Formulas (Circumference)

Practice 33

> **Reminder**
> To compute the circumference of a circle, multiply the diameter times pi ($\pi = 3.14$) or multiply the radius times 2 times pi ($\pi = 3.14$).
>
> $C = \pi d$ or $C = 2\pi r$
>
> 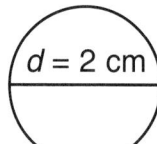 $d = 2$ cm
>
> $C = \pi d$
> $C = 3.14 \times 2$
> $C = 6.28$ cm

Directions: Name the formula, write the equation, and compute the circumference of each circle. The first one is done for you.

1.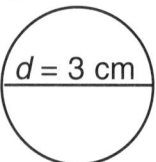

   Formula: $C = \pi d$
   Equation: 3.14 x 3
   Circumference: 9.42 cm

2.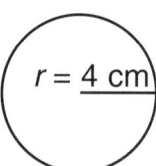

   Formula: $C = 2\pi r$
   Equation: _____
   Circumference: _____

3.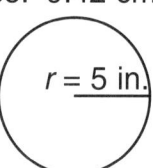

   Formula: _____
   Equation: _____
   Circumference: _____

4.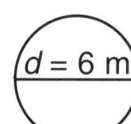

   Formula: _____
   Equation: _____
   Circumference: _____

5.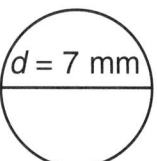

   Formula: _____
   Equation: _____
   Circumference: _____

6.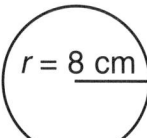

   Formula: _____
   Equation: _____
   Circumference: _____

7.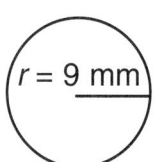

   Formula: _____
   Equation: _____
   Circumference: _____

8.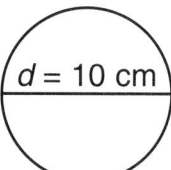

   Formula: _____
   Equation: _____
   Circumference: _____

Practice 34

Directions: Find the missing numbers to complete each function frame.

1. Rule: Output = 3 x input	
Input	Output
3	9
4	12
5	15
6	___
7	___
8	___

2. Rule: Output = Input + 7	
Input	Output
3	10
5	12
7	14
9	___
11	___
13	___

3. Rule: Output = Input – 5	
Input	Output
5	0
7	2
9	4
10	___
13	___
14	___

4. Rule: Output = Input x 4	
Input	Output
5	20
6	24
7	28
8	___
10	___
12	___

5. Rule: Output = Input x 7	
Input	Output
3	21
5	35
6	42
8	___
10	___
11	___
13	___

6. Rule: Output = Input + 6	
Input	Output
3	9
6	12
7	13
9	___
12	___
14	___
17	___

7. Rule: Output = Input x 5	
Input	Output
9	45
11	55
15	75
16	___
18	___
20	___

8. Rule: Output = Input – 7	
Input	Output
14	7
16	9
17	10
19	___
20	___
21	___

Simple Functions

Practice 35

Directions: Find the missing numbers to complete each function frame. Determine the rule for each function.

1. **Rule:** $f(n) = n \times 3$

n	f(n)
1	3
2	6
4	12
5	___
7	___
9	___
11	___

2. **Rule:** $f(n) = $ _____

n	f(n)
7	10
8	11
10	13
11	___
13	___
16	___
18	___

3. **Rule:** $f(n) = $ _____

n	f(n)
2	8
5	20
7	28
8	___
9	___
11	___
14	___

4. **Rule:** $f(n) = $ _____

n	f(n)
2	10
4	12
5	13
6	___
7	___
9	___
10	___

5. **Rule:** $f(n) = $ _____

n	f(n)
5	2
6	3
7	4
9	___
10	___
13	___
16	___

6. **Rule:** $f(n) = $ _____

n	f(n)
11	1
12	2
14	4
17	___
21	___
24	___
23	___

7. **Rule:** $f(n) = $ _____

n	f(n)
3	21
4	28
6	42
7	___
9	___
10	___
12	___

8. **Rule:** $f(n) = $ _____

n	f(n)
22	18
20	16
19	15
16	___
15	___
13	___
10	___

Practice 36

Directions: Determine the missing numbers in each sequence below.

1. (3, 6, 9, 12, 15, _____, _____, _____, _____, _____, _____)

2. (1, 3, 5, 7, 9, _____, _____, _____, _____, _____, _____)

3. (5, 9, 13, 17, 21, _____, _____, _____, _____, _____, _____)

4. (0, 5, 10, 15, 20, _____, _____, _____, _____, _____, _____)

5. (1, 9, 17, 25, 33, _____, _____, _____, _____, _____, _____)

6. (8, 16, 24, 32, _____, _____, _____, _____, _____, _____)

7. (29, 26, 23, 20, _____, _____, _____, _____, _____, _____)

8. (99, 87, 75, 63, 51, _____, _____, _____, _____)

9. (144, 132, 120, 108, _____, _____, _____, _____, _____, _____)

10. (19, 29, 39, 49, _____, 69, _____, _____, _____, _____, _____)

11. (1, 5, 10, 16, 23, 31, _____, _____, _____, _____, _____, _____)

12. (4, 8, 16, 32, 64, _____, _____, _____, _____, _____)

13. (1, 3, 9, 27, 81, _____, _____, _____, _____, _____)

14. (20, 40, 60, _____, 100, _____, _____, _____, _____, _____)

15. (1, 1, 2, 3, 5, 8, 13, 21, _____, _____, _____, _____, _____)

16. (3, 3, 6, 9, 15, 24, 39, _____, _____, _____, _____, _____)

Number Sequences and Algebraic Symbols

Test Practice 1

Directions: Determine the missing number in each space below.

1. 5 + _____ = 42	2. 27 − _____ = 13
(A) 37 (B) 25 (C) 57 (D) 17	(A) 30 (B) 4 (C) 14 (D) 40

3. _____ − 18 = 42	4. _____ + 36 = 47
(A) 60 (B) 24 (C) 58 (D) 14	(A) 11 (B) 13 (C) 12 (D) 83

5. 12 x _____ = 48	6. _____ x 6 = 72
(A) 24 (B) 36 (C) 4 (D) 6	(A) 9 (B) 12 (C) 7 (D) 10

Directions: Determine the value of each letter in the expressions below.

7. $22 + n = 49$	8. $a + 17 = 32$
(A) 11 (B) 27 (C) 71 (D) 37	(A) 15 (B) 49 (C) 13 (D) 14

9. $n \times 9 = 36$	10. $n \times 5 = 55$
(A) 14 (B) 4 (C) 27 (D) 5	(A) 50 (B) 60 (C) 11 (D) 12

Directions: Identify each algebraic symbol.

11. =	12. <
(A) not equal (B) less than (C) greater than (D) equals	(A) not equal (B) less than (C) greater than (D) equals

13. >	14. ≠
(A) not equal (B) less than (C) greater than (D) equals	(A) not equal (B) less than (C) greater than (D) equals

Test Practice 2

Directions: Solve these equations using the axioms of equality.

1. $a - 15 = 28$ (A) 13 (B) 33 (C) 53 (D) 43	2. $t + 21 = 46$ (A) 27 (B) 67 (C) 15 (D) 25
3. $n + 13 = 31$ (A) 18 (B) 28 (C) 44 (D) 9	4. $p - 27 = 42$ (A) 49 (B) 69 (C) 15 (D) 68
5. $n \times 12 = 60$ (A) 12 (B) 5 (C) 72 (D) 48	6. $a \times 9 = 45$ (A) 5 (B) 54 (C) 6 (D) 36
7. $\frac{n}{8} = 9$ (A) 1 (B) 72 (C) 17 (D) 84	8. $\frac{c}{9} = 12$ (A) 3 (B) 96 (C) 212 (D) 108
9. $\frac{t}{7} = 8$ (A) 15 (B) 54 (C) 1 (D) 56	10. $\frac{r}{5} = 15$ (A) 75 (B) 20 (C) 10 (D) 85
11. $b - 17 = 49$ (A) 76 (B) 66 (C) 56 (D) 32	12. $v + 26 = 71$ (A) 35 (B) 55 (C) 45 (D) 97
13. $n \times 13 = 52$ (A) 14 (B) 65 (C) 39 (D) 4	14. $8 \cdot b = 88$ (A) 77 (B) 11 (C) 96 (D) 80

Evaluating Exponents & Expressions

Test Practice 3

Directions: Evaluate these expressions.

1. 5^2 (A) 25 (B) 7 (C) 20 (D) 10	2. 4^3 (A) 12 (B) 20 (C) 64 (D) 7
3. $2^2 + 9$ (A) 13 (B) 14 (C) 36 (D) 31	4. $(29 - 6) + 12$ (A) 36 (B) 35 (C) 47 (D) 11
5. $(19 - 7) + 33$ (A) 35 (B) 45 (C) 21 (D) 59	6. $7^2 + 2^2$ (A) 18 (B) 45 (C) 10 (D) 53
7. $6^2 \times 5$ (A) 41 (B) 60 (C) 180 (D) 160	8. $19 - (6 + 7)$ (A) 20 (B) 6 (C) 16 (D) 7
9. $9^2 + 3^2$ (A) 99 (B) 24 (C) 90 (D) 87	10. $8^2 - 17$ (A) 1 (B) 48 (C) 33 (D) 47
11. $11^2 - 37$ (A) 64 (B) 48 (C) 15 (D) 84	12. $5^3 + 5^2$ (A) 25 (B) 100 (C) 15 (D) 150
13. $(22 - 5) - (2 + 6)$ (A) 9 (B) 19 (C) 21 (D) 7	14. $(19 - 4) - (9 + 3)$ (A) 3 (B) 9 (C) 4 (D) 27
15. $(33 - 9) - (7 + 4)$ (A) 35 (B) 21 (C) 13 (D) 23	16. $(28 - 7) - (16 - 4)$ (A) 35 (B) 9 (C) 1 (D) 33

Order of Operations

Test Practice 4

Directions: Evaluate these expressions using the correct order of operations.

1. 18 ÷ 6 x 4 + 7 (A) 21 (B) 18 (C) 19 (D) 44	2. 8 x 6 − 9 x 5 (A) 3 (B) 195 (C) 19 (D) 4
3. 36 ÷ 9 x 7 − 7 (A) 0 (B) 4 (C) 21 (D) 20	4. 12 + 7 x 3 − 11 (A) 46 (B) 56 (C) 22 (D) 44
5. (17 + 3) − (14 + 5) (A) 1 (B) 39 (C) 11 (D) 2	6. (27 + 3) − (12 x 2) (A) 5 (B) 6 (C) 36 (D) 16
7. 25 − (3 x 7) − 2 (A) 152 (B) 2 (C) 1 (D) 4	8. 19 + (5 x 6) − 9 (A) 40 (B) 50 (C) 30 (D) 135
9. 4^2 + (3 x 8) − 9 (A) 21 (B) 31 (C) 143 (D) 33	10. 7^2 − (7 x 6) (A) 7 (B) 252 (C) 91 (D) 28
11. 8^2 − (5 x 11) + 7 (A) 16 (B) 46 (C) 128 (D) 18	12. (13 x 4) − 2^3 (A) 60 (B) 48 (C) 54 (D) 44
13. (39 − 3^2) + 17 (A) 45 (B) 50 (C) 36 (D) 47	14. 9 x 2^2 − 4 − 8 (A) 25 (B) 40 (C) 24 (D) 6
15. 93 − (4^2 x 5) (A) 23 (B) 13 (C) 385 (D) 3	16. 12 + 5^2 − (2 x 7) (A) 33 (B) 23 (C) 8 (D) 245

© Teacher Created Materials, Inc.   #8635 Practice Makes Perfect: Pre-Algebra

Working with Integers & Variables

Test Practice 5

Directions: Place these integers in the correct order from least to greatest.

1. ⁻6, ⁺7, ⁻4	2. ⁻9, ⁻11, ⁺8
(A) ⁻4, ⁺7, ⁻6	(A) ⁻9, ⁻11, ⁺8
(B) ⁻6, ⁻4, ⁺7	(B) ⁺8, ⁻9, ⁻11
(C) ⁻4, ⁻6, ⁺7	(C) ⁻11, ⁻9, ⁺8
(D) ⁺7, ⁻6, ⁻4	(D) ⁻9, ⁺8, ⁻11
3. 0, ⁻12, ⁺8	4. ⁺9, ⁻9, 0
(A) ⁺8, 0, ⁻12	(A) ⁻9, ⁺9, 0
(B) 0, ⁻12, ⁺8	(B) 0, ⁻9, ⁺9
(C) ⁻12, 0, ⁺8	(C) ⁻9, 0, ⁺9
(D) ⁻12, ⁺8, 0	(D) ⁺9, ⁻9, 0

Directions: Add the integers in each problem below.

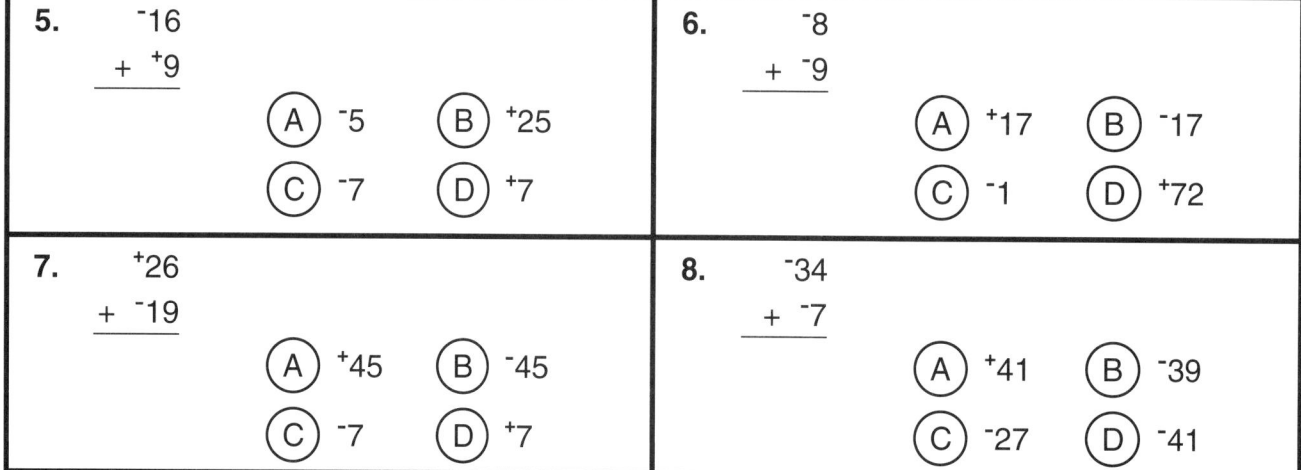

Directions: Evaluate these expressions. ($n = 9$ $b = 4$)

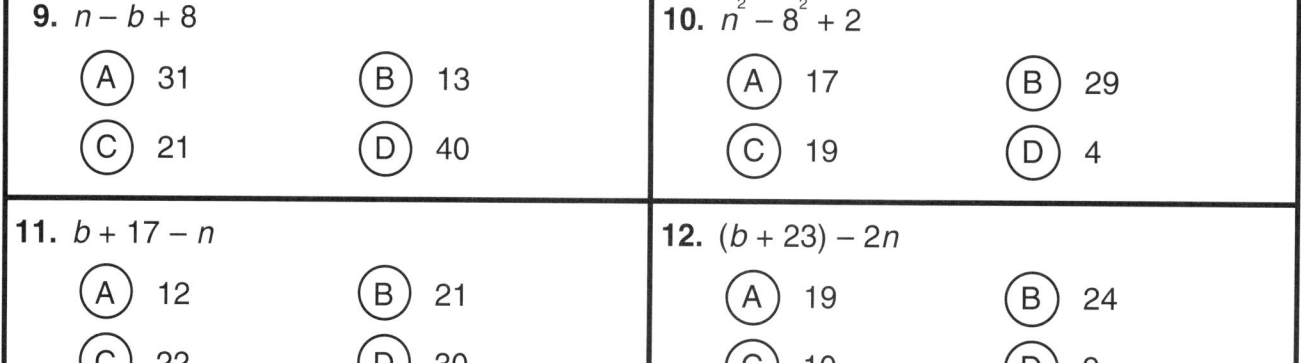

#8635 *Practice Makes Perfect: Pre-Algebra* © Teacher Created Materials, Inc.

Test Practice 6

Directions: Answer each question.

A.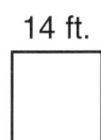
14 ft.

B.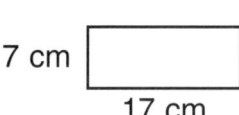
7 cm
17 cm

C.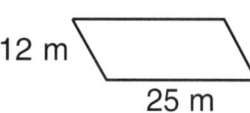
12 m
25 m

D.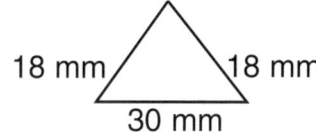
18 mm 18 mm
30 mm

1. What is the perimeter of square A?
   - (A) 56 ft.
   - (B) 140 ft.
   - (C) 6 ft.
   - (D) 5600 ft.

2. What is the perimeter of rectangle B?
   - (A) 49 cm
   - (B) 119 cm
   - (C) 48 cm
   - (D) 24 cm

3. What is the perimeter of parallelogram C?
   - (A) 76 m
   - (B) 37 m
   - (C) 74 m
   - (D) 300 m

4. What is the perimeter of triangle D?
   - (A) 12 mm
   - (B) 66 mm
   - (C) 48 mm
   - (D) 540 mm

5. What is the area of square A?
   - (A) 56 sq. ft.
   - (B) 428 sq. ft.
   - (C) 48 sq. ft.
   - (D) 196 sq. ft.

6. What is the area of rectangle B?
   - (A) 19 sq. cm
   - (B) 119 sq. cm
   - (C) 12 sq. cm
   - (D) 34 sq. cm

7. What is the area of a parallelogram with a base of 25 meters and a height of 9 meters?
   - (A) 34 sq. m
   - (B) 4 sq. m
   - (C) 16 sq. m
   - (D) 225 sq. m

8. What is the circumference of a circle with a diameter of 5 centimeters? ($\pi = 3.14$)
   - (A) 1570 cm
   - (B) 15.7 cm
   - (C) 8.14 cm
   - (D) 16.7 cm

Directions: Find the missing number in each function frame.

9.	n	$f(n)$
	9	27
	10	30
	12	36
	15	___

- (A) 44
- (B) 40
- (C) 45
- (D) 60

10.	n	$f(n)$
	9	17
	11	19
	15	23
	18	___

- (A) 26
- (B) 54
- (C) 16
- (D) 27

Answer Sheet

Test Practice 1	Test Practice 2	Test Practice 3
1. Ⓐ Ⓑ Ⓑ Ⓓ	1. Ⓐ Ⓑ Ⓑ Ⓓ	1. Ⓐ Ⓑ Ⓑ Ⓓ
2. Ⓐ Ⓑ Ⓒ Ⓓ	2. Ⓐ Ⓑ Ⓒ Ⓓ	2. Ⓐ Ⓑ Ⓒ Ⓓ
3. Ⓐ Ⓑ Ⓒ Ⓓ	3. Ⓐ Ⓑ Ⓒ Ⓓ	3. Ⓐ Ⓑ Ⓒ Ⓓ
4. Ⓐ Ⓑ Ⓒ Ⓓ	4. Ⓐ Ⓑ Ⓒ Ⓓ	4. Ⓐ Ⓑ Ⓒ Ⓓ
5. Ⓐ Ⓑ Ⓒ Ⓓ	5. Ⓐ Ⓑ Ⓒ Ⓓ	5. Ⓐ Ⓑ Ⓒ Ⓓ
6. Ⓐ Ⓑ Ⓒ Ⓓ	6. Ⓐ Ⓑ Ⓒ Ⓓ	6. Ⓐ Ⓑ Ⓒ Ⓓ
7. Ⓐ Ⓑ Ⓒ Ⓓ	7. Ⓐ Ⓑ Ⓒ Ⓓ	7. Ⓐ Ⓑ Ⓒ Ⓓ
8. Ⓐ Ⓑ Ⓒ Ⓓ	8. Ⓐ Ⓑ Ⓒ Ⓓ	8. Ⓐ Ⓑ Ⓒ Ⓓ
9. Ⓐ Ⓑ Ⓒ Ⓓ	9. Ⓐ Ⓑ Ⓒ Ⓓ	9. Ⓐ Ⓑ Ⓒ Ⓓ
10. Ⓐ Ⓑ Ⓒ Ⓓ	10. Ⓐ Ⓑ Ⓒ Ⓓ	10. Ⓐ Ⓑ Ⓒ Ⓓ
11. Ⓐ Ⓑ Ⓒ Ⓓ	11. Ⓐ Ⓑ Ⓒ Ⓓ	11. Ⓐ Ⓑ Ⓒ Ⓓ
12. Ⓐ Ⓑ Ⓒ Ⓓ	12. Ⓐ Ⓑ Ⓒ Ⓓ	12. Ⓐ Ⓑ Ⓒ Ⓓ
13. Ⓐ Ⓑ Ⓒ Ⓓ	13. Ⓐ Ⓑ Ⓒ Ⓓ	13. Ⓐ Ⓑ Ⓒ Ⓓ
14. Ⓐ Ⓑ Ⓒ Ⓓ	14. Ⓐ Ⓑ Ⓒ Ⓓ	14. Ⓐ Ⓑ Ⓒ Ⓓ
		15. Ⓐ Ⓑ Ⓒ Ⓓ
		16. Ⓐ Ⓑ Ⓒ Ⓓ

Test Practice 4	Test Practice 5	Test Practice 6
1. Ⓐ Ⓑ Ⓑ Ⓓ	1. Ⓐ Ⓑ Ⓑ Ⓓ	1. Ⓐ Ⓑ Ⓑ Ⓓ
2. Ⓐ Ⓑ Ⓒ Ⓓ	2. Ⓐ Ⓑ Ⓒ Ⓓ	2. Ⓐ Ⓑ Ⓒ Ⓓ
3. Ⓐ Ⓑ Ⓒ Ⓓ	3. Ⓐ Ⓑ Ⓒ Ⓓ	3. Ⓐ Ⓑ Ⓒ Ⓓ
4. Ⓐ Ⓑ Ⓒ Ⓓ	4. Ⓐ Ⓑ Ⓒ Ⓓ	4. Ⓐ Ⓑ Ⓒ Ⓓ
5. Ⓐ Ⓑ Ⓒ Ⓓ	5. Ⓐ Ⓑ Ⓒ Ⓓ	5. Ⓐ Ⓑ Ⓒ Ⓓ
6. Ⓐ Ⓑ Ⓒ Ⓓ	6. Ⓐ Ⓑ Ⓒ Ⓓ	6. Ⓐ Ⓑ Ⓒ Ⓓ
7. Ⓐ Ⓑ Ⓒ Ⓓ	7. Ⓐ Ⓑ Ⓒ Ⓓ	7. Ⓐ Ⓑ Ⓒ Ⓓ
8. Ⓐ Ⓑ Ⓒ Ⓓ	8. Ⓐ Ⓑ Ⓒ Ⓓ	8. Ⓐ Ⓑ Ⓒ Ⓓ
9. Ⓐ Ⓑ Ⓒ Ⓓ	9. Ⓐ Ⓑ Ⓒ Ⓓ	9. Ⓐ Ⓑ Ⓒ Ⓓ
10. Ⓐ Ⓑ Ⓒ Ⓓ	10. Ⓐ Ⓑ Ⓒ Ⓓ	10. Ⓐ Ⓑ Ⓒ Ⓓ
11. Ⓐ Ⓑ Ⓒ Ⓓ	11. Ⓐ Ⓑ Ⓒ Ⓓ	
12. Ⓐ Ⓑ Ⓒ Ⓓ	12. Ⓐ Ⓑ Ⓒ Ⓓ	
13. Ⓐ Ⓑ Ⓒ Ⓓ		
14. Ⓐ Ⓑ Ⓒ Ⓓ		
15. Ⓐ Ⓑ Ⓒ Ⓓ		
16. Ⓐ Ⓑ Ⓒ Ⓓ		

Teacher Created Materials

"Created *by* Teachers *for* Teachers"

Quality Resource Books
- language arts
- social studies
- math
- science
- technology
- the arts

Decorative Products
- 2-sided decorations
- 3-D decorations
- accent dazzlers
- awards
- badges
- banners
- bookmarks
- border trim
- bulletin boards
- file folders
- incentive charts
- name plates
- name tags
- notepads
- pocket folders
- postcards
- stickers

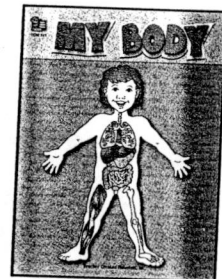

For more information, visit our Web site: www.teachercreated.com

Answer Key

Page 4
1. 35
2. 72
3. 117
4. 111
5. 17
6. 25
7. 37
8. 44
9. 29
10. 39
11. 27
12. 68
13. 77
14. 39
15. 42
16. 49
17. 34
18. 78
19. 171
20. 86
21. 87
22. 23
23. 92
24. 85

Page 5
1. 32
2. 69
3. 60
4. 73
5. 30
6. 35
7. 20
8. 39
9. 37
10. 50
11. 10
12. 29
13. 20
14. 16
15. 60
16. 30
17. 42
18. 24
19. 23
20. 60
21. 50
22. 28
23. 66
24. 80

Page 6
1. 7
2. 8
3. 5
4. 8
5. 12
6. 8
7. 7
8. 8
9. 5
10. 11
11. 10
12. 8
13. 10
14. 12
15. 4
16. 3
17. 6
18. 6
19. 13
20. 4
21. 3
22. 8
23. 10
24. 19

Page 7
1. 19 is greater than 16
2. 42
3. 19 is less than 24
4. 18 is less than 33
5. 120
6. 62
7. 6 is greater than 5
8. 31
9. 22
10. 81 is greater than 72
11. 45 is greater than 36
12. 18 is less than 35
13. 48 is less than 77
14. 8 equals 8
15. 44 is not equal to 40
16. 54 is not equal to 67
17. 104 is greater than 64
18. 9 equals 9
19. 38 is less than 40
20. 27 equals 27

Page 8
1. 37
2. 71
3. 39
4. 38
5. 39
6. 58
7. 10
8. 10
9. 20
10. 70
11. 31
12. 18
13. 16
14. 127
15. 50
16. 36
17. 24
18. 2

Page 9
1. 135
2. 90
3. 81
4. 167
5. 331
6. 106
7. 40
8. 122
9. 104
10. 1
11. 101
12. 192
13. 61
14. 14
15. 56
16. 0
17. 57
18. 48
19. 13
20. 256
21. 183

Page 10
1. $n = 20$
2. $n = 36$
3. $c = 30$
4. $b = 25$
5. $n = 20$
6. $g = 30$
7. $a = 35$
8. $p = 30$
9. $n = 50$
10. $b = 30$
11. $v = 40$
12. $m = 40$
13. $t = 34$
14. $c = 20$
15. $a = 50$
16. $n = 33$
17. $d = 40$
18. $g = 50$

Page 11
1. $n = 10$
2. $a = 10$
3. $c = 10$
4. $d = 19$
5. $n = 2$
6. $c = 7$
7. $d = 33$
8. $m = 14$
9. $t = 22$
10. $a = 44$
11. $s = 40$
12. $x = 18$
13. $p = 10$
14. $n = 23$
15. $a = 66$
16. $t = 75$
17. $x = 34$
18. $n = 52$

Page 12
1. $n = 40$
2. $c = 42$
3. $p = 35$
4. $d = 48$
5. $t = 60$
6. $n = 54$
7. $a = 60$
8. $c = 99$
9. $r = 63$
10. $d = 96$
11. $m = 30$
12. $t = 54$
13. $c = 39$
14. $n = 80$
15. $n = 100$

Page 13
1. $n = 4$
2. $c = 10$
3. $a = 4$
4. $n = 6$
5. $s = 9$
6. $t = 9$
7. $n = 9$
8. $p = 7$
9. $w = 10$
10. $c = 7$
11. $z = 4$
12. $b = 3$
13. $m = 6$
14. $d = 7$
15. $d = 11$
16. $r = 3$
17. $c = 6$
18. $c = 4$

Page 14
1. $n = 14$
2. $d = 5$
3. $q = 8$
4. $c = 41$
5. $m = 4$
6. $n = 8$
7. $v = 108$
8. $n = 56$
9. $n = 20$
10. $t = 23$
11. $n = 6$
12. $a = 57$
13. $d = 4$
14. $c = 7$
15. $d = 42$
16. $n = 55$
17. $t = 44$
18. $s = 24$
19. $s = 11$
20. $d = 3$
21. $w = 63$
22. $t = 48$
23. $b = 48$
24. $a = 144$

Page 15
1. $n = 57$
2. $z = 6$
3. $s = 29$
4. $d = 80$
5. $t = 7$
6. $b = 57$
7. $n = 36$
8. $r = 132$
9. $t = 85$
10. $v = 20$
11. $c = 12$
12. $b = 104$
13. $c = 63$
14. $t = 72$
15. $v = 81$
16. $n = 81$
17. $n = 10$
18. $c = 13$
19. $n = 28$
20. $q = 4$
21. $p = 62$
22. $r = 80$
23. $d = 99$
24. $n = 11$

Page 16
1. $n = 5$
2. $p = 17$
3. $a = 22$
4. $r = 7$
5. $n = 28$
6. $c = 16$
7. $b = 9$
8. $c = 8$
9. $q = 12$
10. $c = 16$
11. $r = 4$
12. $b = 9$
13. $t = 8$
14. $c = 15$
15. $r = 1$
16. $c = 20$
17. $n = 13$
18. $r = 7$

Page 17
1. 14
2. 25
3. 15
4. 7
5. 6
6. 16
7. 23
8. 12
9. 3
10. 21
11. 7
12. 12
13. 6
14. 1
15. 12
16. 5
17. 49
18. 12
19. 5
20. 22
21. 19
22. 124
23. 7
24. 48
25. 3
26. 17
27. 21
28. 66
29. 8
30. 1

Page 18
1. 26
2. 9
3. 2
4. 20
5. 7
6. 6
7. 17
8. 28
9. 51
10. 1
11. 10
12. 6
13. 4
14. 26
15. 6
16. 13
17. 3
18. 8
19. 17
20. 3
21. 4
22. 5
23. 10
24. 23
25. 14
26. 21
27. 5
28. 7
29. 7
30. 16

Page 19
1. 36
2. 9
3. 25
4. 4
5. 100
6. 121
7. 125
8. 216
9. 81
10. 512
11. 49
12. 1,000
13. 343
14. 169
15. 8
16. 81
17. 16
18. 729
19. 125
20. 32
21. 196
22. 1
23. 225
24. 400
25. 900
26. 1,600
27. 2,500

Page 20
1. 13
2. 8
3. 22
4. 6
5. 2
6. 7
7. 4
8. 60
9. 10
10. 61
11. 5
12. 64
13. 4
14. 13
15. 53
16. 17
17. 58
18. 50

Page 21
1. 36
2. 20
3. 63
4. 225
5. 44
6. 351
7. 135
8. 196
9. 875
10. 441
11. 1,573
12. 539
13. 200
14. 392
15. 833
16. 275
17. 440
18. 140

Page 22
1. 8
2. 22
3. 9
4. 8
5. 19
6. 13
7. 4
8. 18

© Teacher Created Materials, Inc. #8635 Practice Makes Perfect: Pre-Algebra 47

Answer Key (cont.)

9. 18
10. 15
11. 25
12. 21
13. 54
14. 55
15. 51
16. 4
17. 32
18. 34

Page 23
1. 19
2. 7
3. 21
4. 27
5. 41
6. 54
7. 68
8. 59
9. 5
10. 63
11. 14
12. 53
13. 46
14. 7
15. 1
16. 27
17. 7
18. 65
19. 56
20. 2
21. 1

Page 24
1. 14
2. 21
3. 11
4. 59
5. 85
6. 53
7. 88
8. 22
9. 3
10. 68
11. 4
12. 32
13. 54
14. 24
15. 1
16. 40
17. 5
18. 49
19. 1
20. 6
21. 1
22. 12
23. 19
24. 1

Page 25
1. 10
2. 30
3. 25
4. 1
5. 63
6. 53
7. 6
8. 18
9. 0
10. 20
11. 86
12. 36
13. 8
14. 5
15. 18

Page 26
1. 5
2. 11
3. 13
4. 16
5. 33
6. 1
7. 5
8. 24
9. 40
10. 28
11. 340
12. 124
13. 27
14. 7
15. 28

Page 27
1. 17
2. 1
3. 11
4. 23
5. 1
6. 3
7. 24
8. 18
9. 42
10. 13
11. 32
12. 9
13. 18
14. 32
15. 20
16. 2
17. 0
18. 9

Page 28
1. 5
2. 10
3. 11
4. 22
5. 64
6. 10
7. 14
8. 127
9. 62
10. 0
11. 57
12. 174
13. 32
14. 22
15. 32
16. 16
17. 6
18. 167

Page 29
1. ⁻2, 0, ⁺1
2. ⁻6, ⁻5, ⁺4
3. ⁻9, 0, ⁺6
4. ⁻9, ⁻4, ⁻3
5. ⁻8, ⁻6, ⁺8
6. ⁻6, 0, ⁺7
7. ⁻5, ⁻1, ⁺1
8. ⁻7, 0, ⁺9
9. ⁻9, 0, ⁺6
10. ⁻6, ⁻1, ⁺3
11. ⁻12, ⁻6, ⁺5
12. ⁻7, ⁻5, ⁺6
13. ⁻10, ⁻6, ⁻3
14. ⁻9, ⁻5, 0
15. ⁻13, ⁻8, ⁻5, 0
16. ⁻7, ⁻6, ⁺8, ⁺10
17. ⁻20, ⁻14, ⁻6, ⁺7
18. ⁻12, ⁻7, 0, ⁺11
19. ⁻21, ⁻16, ⁻7, ⁺7
20. ⁻15, ⁻12, 0, ⁺12
21. ⁻17, ⁻9, ⁻5, ⁻4, 0
22. ⁻6, ⁻5, ⁻1, ⁺5, ⁺7

Page 30
1. ⁺2
2. ⁻3
3. ⁻17
4. ⁻18
5. ⁻2
6. ⁻3
7. ⁻3
8. ⁻32
9. ⁺2
10. ⁺5
11. ⁻4
12. ⁺8
13. ⁻36
14. ⁺1
15. ⁻24
16. ⁻10
17. ⁺7
18. ⁻9
19. ⁻9
20. ⁺7
21. ⁻38
22. ⁻16
23. ⁻19
24. ⁺14
25. ⁺2
26. ⁻27
27. ⁺4

Page 31
1. ⁺1
2. ⁻15
3. ⁻13
4. ⁻10
5. ⁻19
6. ⁻21
7. ⁺6
8. ⁺10
9. ⁻26
10. ⁻15
11. ⁻3
12. ⁺8
13. ⁻5
14. ⁻20
15. ⁻38
16. ⁻23
17. ⁻20
18. ⁺21
19. ⁺29
20. ⁺8
21. ⁻5
22. ⁻15

Page 32
1. 20 cm
2. 26 cm
3. 24 in.
4. 58 in.
5. 52 mm
6. 84 mm
7. 56 cm
8. 104 in.
9. 132 m
10. 102 in.

Page 33
1. 15 cm
2. 42 cm
3. 64 in.
4. 58 in.
5. 57 mm
6. 66 cm
7. 132 cm
8. 164 in.
9. 150 m
10. 90 in.

Page 34
1. 126 sq. cm
2. 1,000 sq. mm
3. 529 sq. in.
4. 352 sq. cm
5. 1,521 sq. in.
6. 957 sq. in.
7. 1,036 sq. mm
8. 748 sq. ft.
9. 625 sq. cm
10. 384 sq. cm

Page 35
1. 120 sq. cm
2. 247 sq. cm
3. 150 sq. m
4. 216 sq. in.
5. 1,280 sq. mm
6. 425 sq. cm
7. 2,000 sq. cm
8. 204 sq. m
9. 814 sq. in.
10. 234 sq. ft.

Page 36
1. $C = \pi d$
   $C = 3.14 \times 3$
   $C = 9.42$ cm
2. $C = 2\pi r$
   $C = 2 \times 3.14 \times 4$
   $C = 25.12$ cm
3. $C = 2\pi r$
   $C = 2 \times 3.14 \times 5$
   $C = 31.4$ in.
4. $C = \pi d$
   $C = 3.14 \times 6$
   $C = 18.84$ m
5. $C = \pi d$
   $C = 3.14 \times 7$
   $C = 21.98$ mm
6. $C = 2\pi r$
   $C = 2 \times 3.14 \times 8$
   $C = 50.24$ cm
7. $C = 2\pi r$
   $C = 2 \times 3.14 \times 9$
   $C = 56.52$ mm
8. $C = \pi d$
   $C = 3.14 \times 10$
   $C = 31.4$ cm

Page 37
1. 18, 21, 24
2. 16, 18, 20
3. 5, 8, 9
4. 32, 40, 48
5. 56, 70, 77, 91
6. 15, 18, 20, 23
7. 80, 90, 100
8. 12, 13, 14

Page 38
1. $n \times 3$; 15, 21, 27, 33
2. $n + 3$; 14, 16, 19, 21
3. $n \times 4$; 32, 36, 44, 56
4. $n + 8$; 14, 15, 17, 18
5. $n - 3$; 6, 7, 10, 13
6. $n - 10$; 7, 11, 14, 13
7. $n \times 7$; 49, 63, 70, 84
8. $n - 4$; 12, 11, 9, 6

Page 39
1. 18, 21, 24, 27, 30, 33
2. 11, 13, 15, 17, 19, 21
3. 25, 29, 33, 37, 41, 45
4. 25, 30, 35, 40, 45, 50
5. 41, 49, 57, 65, 73, 81
6. 40, 48, 56, 64, 72, 80
7. 17, 14, 11, 8, 5, 2
8. 39, 27, 15, 3
9. 96, 84, 72, 60, 48, 36
10. 59, 79, 89, 99, 109, 119
11. 40, 50, 61, 73, 86, 100
12. 128; 256; 512; 1,024; 2,048
13. 243; 729; 2,187; 6,561; 19,683
14. 80, 120, 140, 160, 180, 200
15. 34, 55, 89, 144, 233
16. 63, 102, 165, 267, 432

Page 40
1. A
2. C
3. A
4. A
5. C
6. B
7. B
8. A
9. B
10. C
11. D
12. B
13. C
14. A

Page 41
1. D
2. D
3. A
4. B
5. B
6. A
7. B
8. D
9. D
10. A
11. B
12. C
13. D
14. B

Page 42
1. A
2. C
3. A
4. B
5. B
6. D
7. C
8. B
9. C
10. D
11. D
12. D
13. A
14. A
15. C
16. B

Page 43
1. C
2. A
3. C
4. C
5. A
6. B
7. B
8. A
9. B
10. A
11. A
12. D
13. D
14. C
15. B
16. B

Page 44
1. B
2. C
3. C
4. C
5. C
6. B
7. D
8. D
9. B
10. C
11. A
12. D

Page 45
1. A
2. C
3. C
4. B
5. D
6. B
7. D
8. B
9. C
10. A